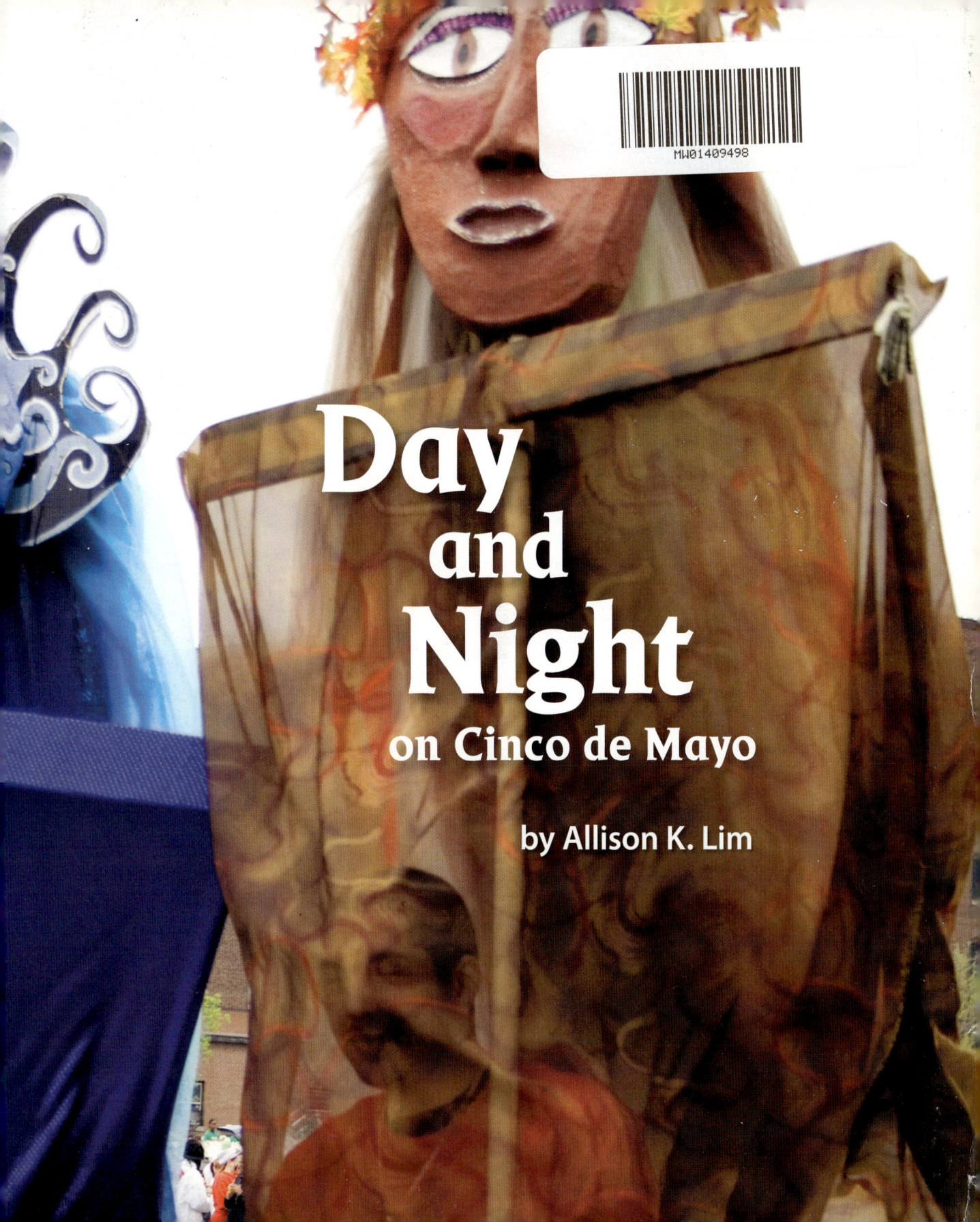

# Day and Night
## on Cinco de Mayo

by Allison K. Lim

# Contents

Science Vocabulary . . . . . . 4

Cinco de Mayo . . . . . . . . 8

Day and Night in Mexico . . . 12

The Day Sky . . . . . . . . . 14

The Night Sky . . . . . . . . 18

Phases of the Moon . . . . . 20

Conclusion . . . . . . . . . . 28

Share and Compare . . . . . 29

Science Career . . . . . . . . 30

Index . . . . . . . . . . . . . 32

3

# Science Vocabulary

**sun**
The **sun** is the star that is nearest to Earth.

The **sun** gives light to Earth.

**moon**
The **moon** is the brightest object in the sky at night.

The **moon** is easier to see at night than it is during the day.

**star**
A **star** is an object in the sky that gives off light.

On a clear night, you can see **stars** in the sky.

**phase**
A **phase** is how the moon looks from Earth.

The moon looks different as it goes through its **phases.**

**telescope**
A **telescope** is a tool that makes objects in the sky look bigger and closer.

A **telescope** can help you see details on the moon.

## My Science Vocabulary

| crater |
| moon |
| phase |
| shadow |
| star |
| sun |
| telescope |

### crater
A **crater** is a dent on the moon's surface.

crater

There are many **craters** on the moon.

### shadow
A **shadow** is a dark shape made when an object blocks light.

There are many **shadows** on a sunny day.

7

# Cinco de Mayo

*Cinco de Mayo* is a holiday.

*Cinco de Mayo* is on May 5.

People celebrate it in Mexico and also in the United States.

People have a *fiesta*.

A fiesta is a party.

The fiesta starts in the morning.
It lasts until late at night.

# Day and Night in Mexico

Day and night happen every 24 hours. The sky is usually light in the day.

Sometimes the moon can be seen in the daytime.

The sky changes color.
The sky is usually dark at night.

# The Day Sky

The **sun** rises.

It lights Earth.

It's Cinco de Mayo!

**sun**

The **sun** is the star that is nearest to Earth.

Cinco de Mayo begins with a parade. People dance and sing.

Dancers block the sun's light.

The dancers make **shadows** on the ground.

**shadow**

A **shadow** is a dark shape made when an object blocks light.

17

# The Night Sky

At night, the sky gets dark. Sometimes, people see **stars** in the night sky.

**star**

A **star** is an object in the sky that gives off light.

They might see the **moon**, too.

**moon**

The **moon** is the brightest object in the sky at night.

# Phases of the Moon

The shape of the moon looks like it changes.
But it doesn't change.

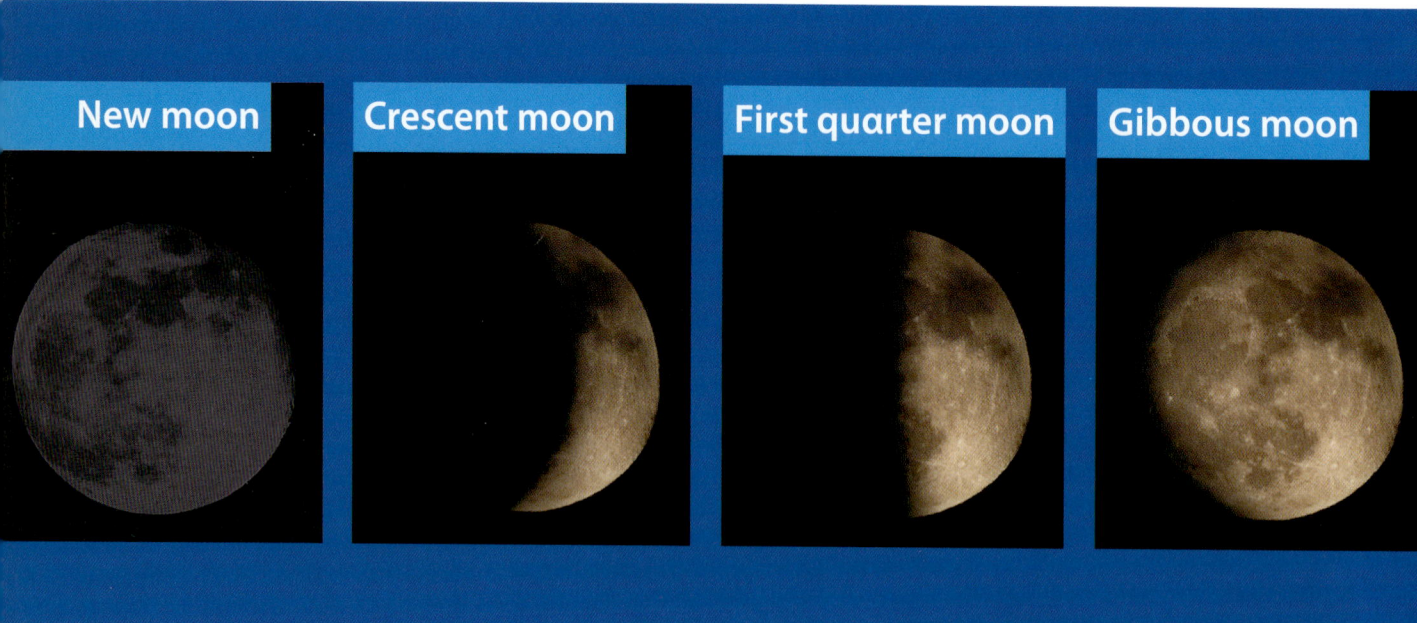

20

These are the **phases** of the moon.

**phase**

A **phase** is how the moon looks from Earth.

21

People can look at the night sky.

A crescent moon hangs in the sky.

People can also look at the sky through a **telescope**. What might they see?

**telescope**

A **telescope** is a tool that makes objects in the sky look bigger and closer.

People might see the stars and the moon.

# The moon has **craters**.

crater

**crater**

A **crater** is a dent on the moon's surface.

25

What else might be in the night sky?

# Fireworks for Cinco de Mayo!

# Conclusion

During Cinco de Mayo, the sun may light the sky for the parade. At night, the moon and stars may be seen. Fireworks may light the night sky.

## Think About the Big Ideas

1. Compare what you see in the day and night sky.
2. What would Earth be like without the sun?
3. Describe phases of the moon.

# Share and Compare

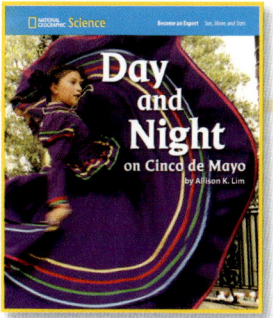 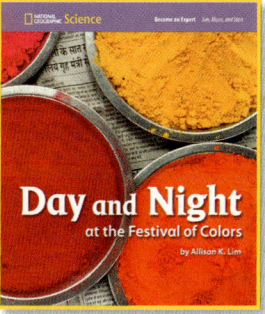 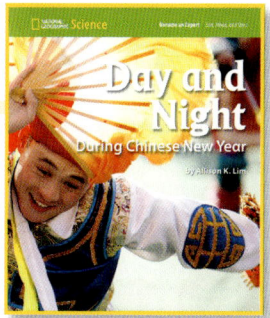

### Turn and Talk

Compare what the sky looks like during the day and at night. How is it the same? How is it different?

### Read

Read your favorite page to a classmate.

### Write

Write about three things you might see in the sky. Share your writing with a classmate.

### Draw

Draw two phases of the moon. Share your drawing with a classmate.

## NATIONAL GEOGRAPHIC Science Career

### Meet Arlin Crotts

Arlin Crotts is a scientist and a teacher. He studies the sun, moon, and stars.

Arlin Crotts heard about flashes of light on the moon. He wants to know what causes these flashes. So he studies the moon to learn more.

He uses special telescopes. They record flashes on the moon. Then he teaches others what he learns.

# Index

**crater** . . . . . . . . . . . . . . . . . . . . 7, 25

**fiesta** . . . . . . . . . . . . . . . . . . . . 10–11

**moon** . . . . . 4, 6–7, 19–21, 24–25, 28–30

**parade** . . . . . . . . . . . . . . . . . . . 15, 28

**phase** . . . . . . . . . . . . . . . 6–7, 21, 29

**shadow** . . . . . . . . . . . . . . . . . . . 7, 17

**star** . . . . . . . . . . . . . 5, 7, 18, 24, 28, 30

**sun** . . . . . . . . . . . . . 4, 7, 14, 16, 28, 30

**telescope** . . . . . . . . . . . 6–7, 23, 30

**Acknowledgments**
Grateful acknowledgment is given to the authors, artists, photographers, museums, publishers, and agents for permission to reprint copyrighted material. Every effort has been made to secure the appropriate permission. If any omissions have been made or if corrections are required, please contact the Publisher.

**Photographic Credits:**
Cover (bg) Robert Galbraith/Reuters/Corbis; Cvr Flap (t), 6 (b), 23 Mark Thiessen and Becky Hale, National Geographic Photographers; Cvr Flap (c), 4 (t), 14 Grigory Kubatyan/Shutterstock; Cvr Flap (b), 7 (t), 25 (inset) DigitalStock/Corbis; Title (bg) Steve Skjold/Alamy Images; 2-3, 7 (b), 16–17 Lawrence Migdale/Lawrence Migdale Photography; 4 (b), 19 Prisma/SuperStock; 5, 18 Giovanni Benintende/Shutterstock; 6 (t), 20–21 David Scheuber/Shutterstock; 8-9 David Young-Wolff/PhotoEdit; 10 Chuck Place/Alamy Images; 11 Dennis MacDonald/PhotoEdit; 12 Ralph Hopkins/Lonely Planet Images; 13, 28 A. Paul Jenkin/Animals Animals; 15 Chuck Place/Alamy Images; 22 Heeb Photos/eStock Photo; 24–25 NASA Johnson Space Center-Earth Sciences and Image Analysis (NASA-JSC-ES&IA); 26–27 Angel Terry/Alamy Images; 30–31 Ira Block; Inside Back Cover (bg) Robert Galbraith/Reuters/Corbis.

Neither the Publisher nor the authors shall be liable for any damage that may be caused or sustained or result from conducting any of the activities in this publication without specifically following instructions, undertaking the activities without proper supervision, or failing to comply with the cautions contained herein.

**Published by National Geographic School Publishing & Hampton-Brown**
Sheron Long, Chairman
Samuel Gesumaria, Vice-Chairman
Alison Wagner, President and CEO
Susan Schaffrath, Executive Vice President, Product Development

**Editorial:** Fawn Bailey, Joseph Baron, Carl Benoit, Jennifer Cocson, Francis Downey, Richard Easby, Mary Clare Goller, Chris Jaeggi, Carol Kotlarczyk, Kathleen Lally, Henry Layne, Allison Lim, Taunya Nesin, Paul Osborn, Chris Siegel, Sara Turner, Lara Winegar, Barbara Wood

**Art, Design, and Production:** Andrea Cockrum, Kim Cockrum, Adriana Cordero, Darius Detwiler, Alicia DiPiero, David Dumo, Jean Elam, Jeri Gibson, Shanin Glenn, Raymond Godfrey, Raymond Hoffmeyer, Rick Holcomb, Cynthia Lee, Anna Matras, Gordon McAlpin, Melina Meltzer, Rick Morrison, Cindy Olson, Christiana Overman, Andrea Pastrano-Tamez, Sean Philpotts, Leonard Pierce, Cathy Revers, Stephanie Rice, Christopher Roy, Janet Sandbach, Susan Scheuer, Margaret Sidlosky, Jonni Stains, Shane Tackett, Andrea Thompson, Andrea Troxel, Ana Vela, Teri Wilson, Brown Publishing Network, Chaos Factory, Inc., Feldman and Associates, Inc.

**The National Geographic Society**
John M. Fahey, Jr., President & Chief Executive Officer
Gilbert M. Grosvenor, Chairman of the Board

**Manufacturing and Quality Management, The National Geographic Society**
Christoper A. Liedel, Chief Financial Officer
George Bounelis, Vice President

Copyright © 2010 The Hampton-Brown Company, Inc., a wholly owned subsidiary of the National Geographic Society, publishing under the imprints National Geographic School Publishing and Hampton-Brown.

All rights reserved. No part of this book may be reproduced or transmitted in any form or by any means, electronic or mechanical, including photocopying, recording, or by an information storage and retrieval system, without permission in writing from the Publisher.

National Geographic and the Yellow Border are registered trademarks of the National Geographic Society.

National Geographic School Publishing
Hampton-Brown
P.O. Box 223220
Carmel, California 93922
www.NGSP.com

Printed in the USA.

ISBN: 978-0-7362-5588-2

10 11 12 13 14 15 16 17

10 9 8 7 6 5 4 3 2 1